PNEUMATIC VALVES FOR AUTOMATION OF IRRIGATION SYSTEMS

PNEUMATIC VALVES
FOR AUTOMATION OF IRRIGATION SYSTEMS

H. R. HAISE,
E. G. KRUSE,
N. A. DIMICK

ISBN: 978-93-90314-33-1

Published: 1965

LECTOR HOUSE LLP
E-MAIL: lectorpublishing@gmail.com

PNEUMATIC VALVES FOR AUTOMATION OF IRRIGATION SYSTEMS

BY

H. R. HAISE, E. G. KRUSE, AND N. A. DIMICK,

Soil and Water Conservation Research Division, Agricultural Research Service

1965

JULY 1965
ARS 41 -104
AGRICULTURAL RESEARCH SERVICE
UNITED STATES DEPARTMENT OF AGRICULTURE

CONTENTS

LIST OF FIGURES

PNEUMATIC VALVES FOR AUTOMATION OF IRRIGATION SYSTEMS[1]

Competition for water among domestic, industrial, recreational, and agricultural users in the Western United States is increasing at an alarming rate. Some authorities believe that irrigated agriculture, one of the low efficiency users of water, could be forced to yield part of its water rights for other purposes unless conservation practices are adopted.

A major cause for inefficient use of water, particularly with surface irrigation systems, is the high labor cost coupled with inexpensive water. Farmers are reluctant to use additional labor just to conserve water. Rather, the farmer often adjusts water application time to labor patterns dictated by general farm operations. The 12-hour irrigation set used by many farmers often results in excessive percolation and runoff losses, particularly where soils are coarser textured and intake rates are high. Furthermore, excessive use of water usually leads to drainage and salinity problems that are costly to alleviate.

Automation of surface irrigation systems offers one way to conserve water on many farms, for the farmer who automates a well-designed irrigation system to save labor will automatically save water. The trend toward the use of expensive sprinkler irrigation systems with a high degree of automation is primarily due to the farmer's desire to reduce labor costs.[2] In Colorado alone, about 110 sprinkler systems capable of irrigating 135 to 140 acres in one circle without attention from the irrigator are currently in use. This development has occurred within the last decade.

Research developments to automate surface irrigation systems have lagged. Bondurant and Humpherys[3] have developed a mechanical gate to apply water sequentially from one bordered area to another. The timing device to open the normally closed gate consists, of a small flow of water that accumulates in a container fastened to the downstream side of the gate. When enough water has accumulated, the trip mechanism releases and the spring-loaded gate swings, upward allowing the water to flow on down the ditch. Auxiliary gates are used to control flow into border strips or furrows. Several other devices have been developed by

---

[1] Space in Foothills Hydraulic Laboratory for model studies provided by Colorado State University, Ft. Collins, Colo.

[2] Bondurant, J. A. Mechanization of surface irrigation. Crops and Soils 14 (1): 23. 1961.

[3] Bondurant, J. A., and Humpherys, A. S. Surface irrigation through automatic control. Agr. Engin.43:20-21, 35. 1962.

farmers,[4] including adaptations of spring-wound clocks to release canvas check dams, overflowing a contour ditch by pulling a canvas dam behind a tractor automatically guided down a water supply ditch, and other similar innovations.

This report describes the development and use of several pneumatic valves capable of being operated by remote control for automatic application of water to graded border strips, to level or nearly level basins, and to furrows supplied from pipeline or open ditch distribution systems.

GENERAL DESCRIPTION

Basically the automated system consists of (1) a pneumatic closure, (2) a three-way solenoid control valve that permits the flow of air into or out of the pneumatic closure, (3) a source of air pressure to inflate the closure, and (4) a centrally located remote control system with timing device to actuate the three-way solenoid control valve by means of a signal transmitted by radio or carried by wire.

LABORATORY INVESTIGATIONS

Pneumatic Valve for Pipe Irrigation System

Irrigation water from underground pipeline distribution systems is released through openings commonly called alfalfa valves, as shown in figure 1. The rate of discharge is controlled by setting the screw-mounted lid to the desired opening.

The pneumatic irrigation valve, illustrated in figure 2, is basically an inflatable O-ring mounted between the alfalfa valve seat and valve lid. It is held in position by a centrally located metal sleeve capable of sliding up and down the threaded screw supporting the valve lid. In an open position the pneumatic valve is forced against the bottom of the valve lid and appears to "ride" on top of the water flowing from the alfalfa valve, and in an inflated or closed position the pneumatic valve forms an annular seal against the valve seat and valve lid, as shown in figure 3. Plastic tubing is used to connect the valve stem with the control valve and pressure source. The alfalfa valve can be preset to any desired opening within the expansion capabilities of the pneumatic valve to regulate flow of water. The pressure required for closure depends on the pressure head in the pipeline system at the point of water delivery. Normally 5 to 10 p.s.i. is adequate, but as much as 25 p.s.i. may be required under extreme operating conditions.

The time required for opening and closing the pneumatic valve is about 15 seconds utilizing present components. Closure time could be reduced by increasing the size of inlet and exhaust ports of the solenoid control valve and eliminating restrictions elsewhere in the air supply system. Actually 15 seconds is a relatively short time when compared with the time water is applied to the land.

The pneumatic valve presently developed can withstand pressures up to about 15 p.s.i. Greater pressures are possible by increasing the strength of the nylon in the nylon-impregnated butyl rubber cover that surrounds the flat natural rubber inner tube. In a laboratory test the butyl covers have withstood abrasive

[4] Pair, C. H. Automation near in irrigation. Agr. Engin. 42: 608-610, 621. 1961.

action caused by more than 300 cycles of opening and closing the valve, which is equivalent to about 50 years of normal field operation. It is known that butyl rubber alone will last at least 15 years under adverse test conditions, including direct exposure to the sun or burying in a compost pile or soil. The material appears ideally suited for the pneumatic valve and is expected to have long life under most operating conditions.

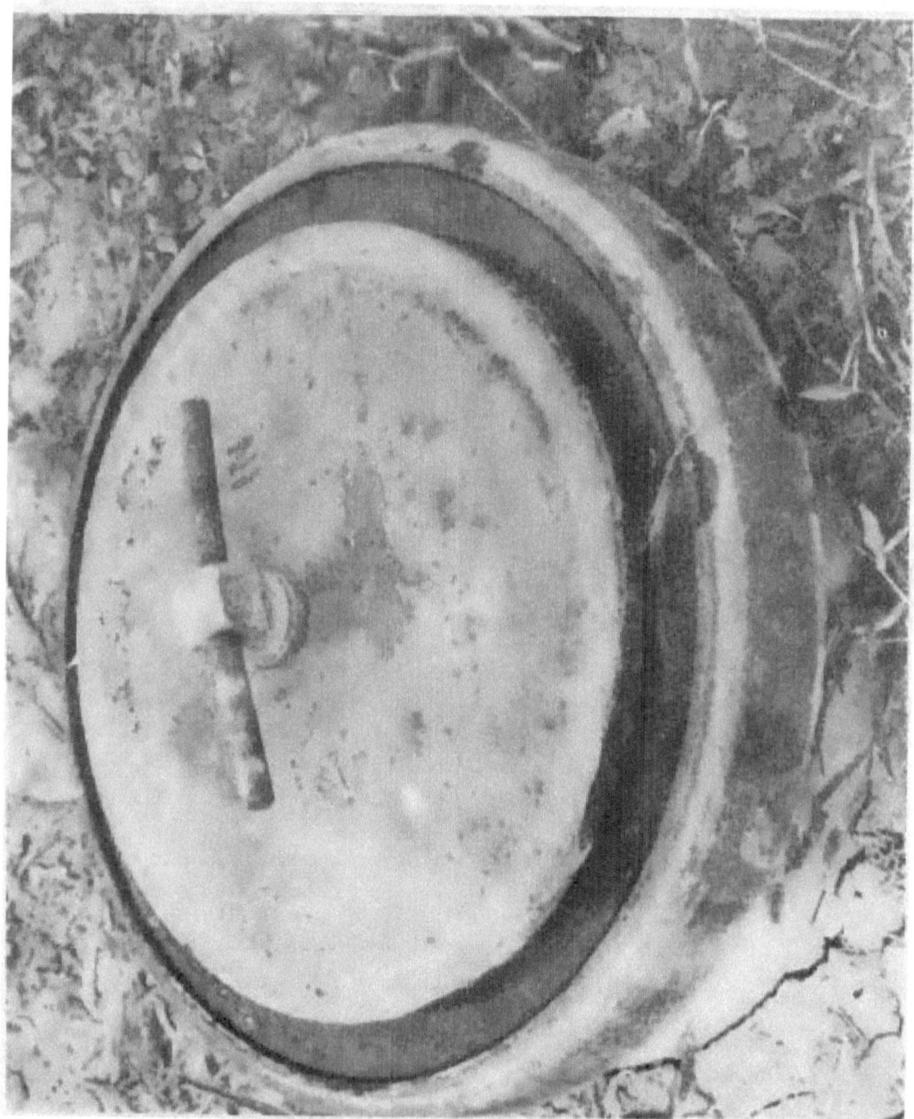

Figure 1.—Twelve-inch alfalfa valve commonly used to apply irrigation water to irrigated fields from buried pipelines.

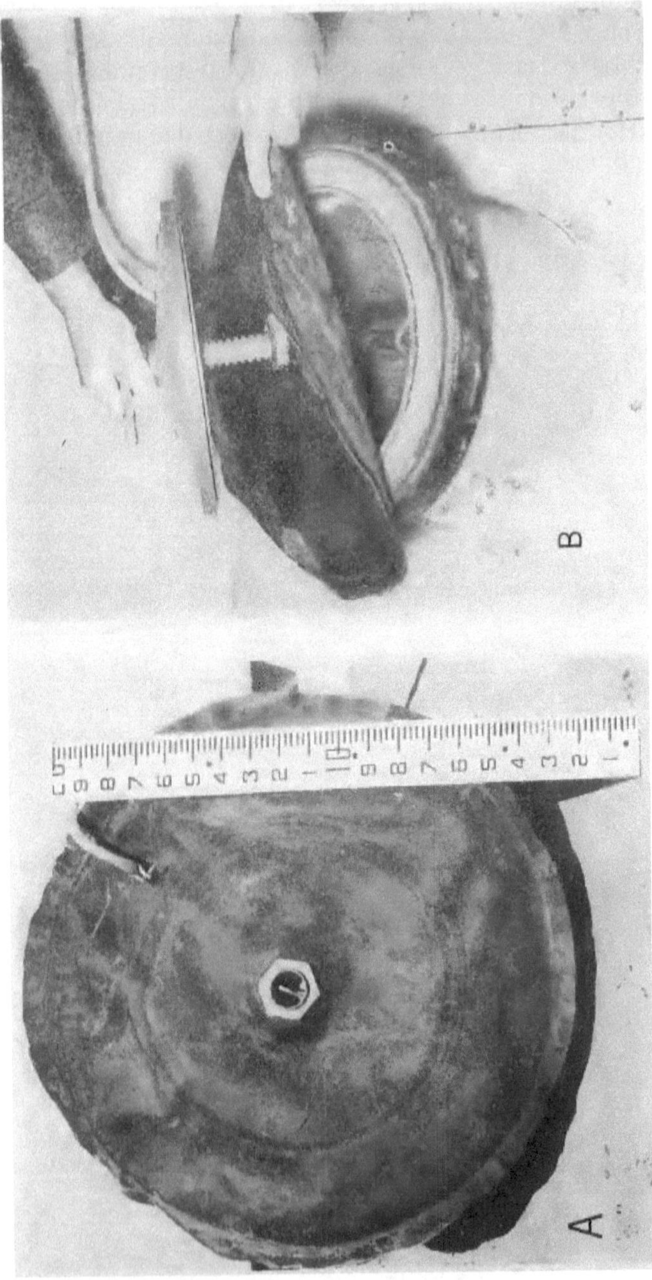

Figure 2.—A, pneumatic 12-inch irrigation valve is basically an inflatable O-ring with a natural rubber inner tube and butyl rubber outer cover. B, valve is being installed on an alfalfa valve used in underground pipeline distribution systems. Pneumatic valve is held in position by center clamp.

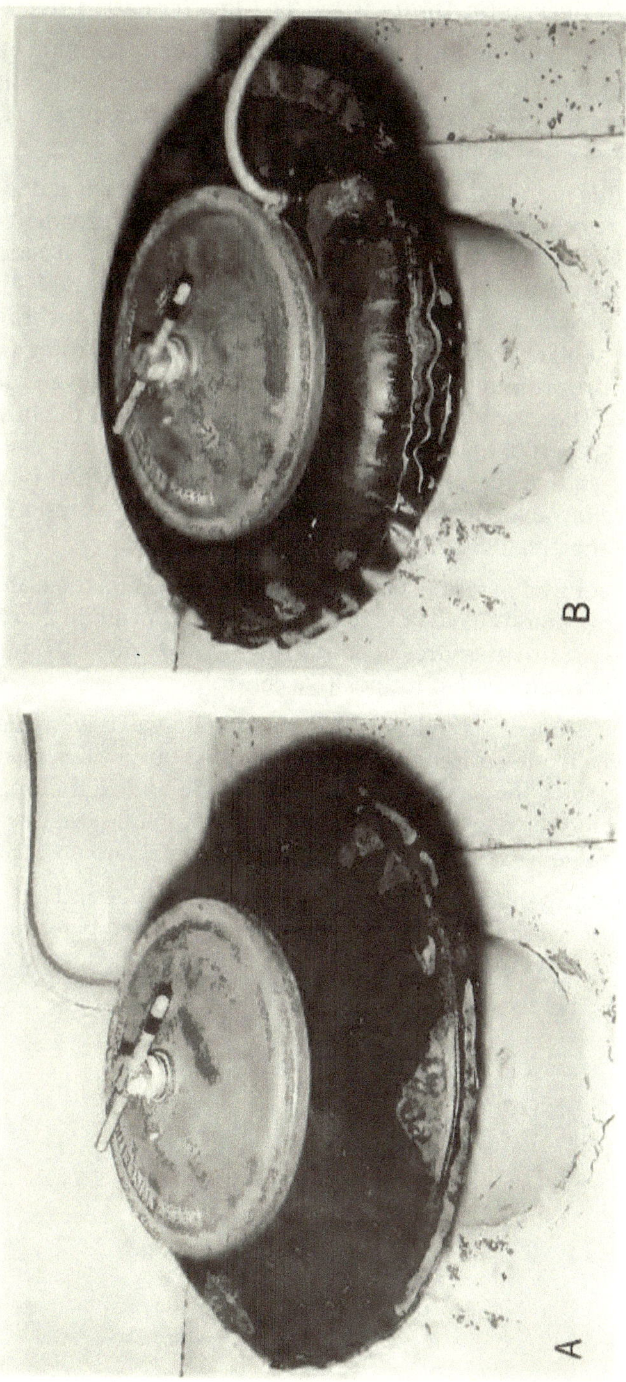

Figure 3. — Pneumatic 12-inch irrigation valve for pipeline deflated or in open position (A) and inflated or in closed position (B).

Modified Pneumatic Alfalfa Valve (O-Ring) for Farm Ditch Turnout

In many lined and unlined field irrigation laterals, permanent turnouts are installed by placing a length of pipe through the ditchbank with a slide gate on the upstream end of the pipe to regulate flow (fig. 4). Concrete pipe in 12-and 15-inch sizes is commonly used in these turnouts.

Basically, the modified 12-inch valve is an O-ring constructed to allow flow of water through a center opening (fig. 5). Metal components are identical to those of the alfalfa valve previously described except for the larger lid (18-inch disk blade). When closed or inflated, the O-ring forms a seal between the valve lid and the valve seat. The lower outer edge of the pneumatic valve is anchored to the base of the turnout at eight positions to prevent the valve from turning up at the edges or from being drawn into the center opening by the force of water flowing over it when deflated. The inside diameter of the O-ring is equal to the turnout diameter to prevent restriction of flow. The valve lid is 6 inches larger in diameter than the valve opening so that the seal is centered on the O-ring when inflated. Like the pneumatic alfalfa valve, the modified turnout valve can be preset to regulate discharge within the inflatable limits of the pneumatic O-ring.

Pressure required to form a watertight seal depends on the water depth in the ditch. Most farm laterals have flow depths ranging from about 2 to 6 feet and thus would require closure pressures of about 1 to 3 p.s.i. A slightly higher pressure or about 5 p.s.i. is recommended for positive control.

Several experimental models of the pneumatic valve were constructed and tested. One was made entirely of nylon-reinforced butyl rubber with vulcanized seams. Spring steel "fingers" were fastened inside and to the upper half of the O-ring to hold the valve in an open position when water flowed over it. Otherwise, the O-ring or valve was drawn into the stream of water and obstructed flow.

The most recent model has a natural rubber inner tube and a nylon-reinforced butyl rubber outer tube with spring steel fingers fastened between the outer and inner tube. The butyl rubber outer tube is stitched at the edges to form the enclosure.

Figure 4.—Open ditch turnout in unlined farm lateral. Depending on the ditch gradient, check structures such as shown in the background may be necessary.

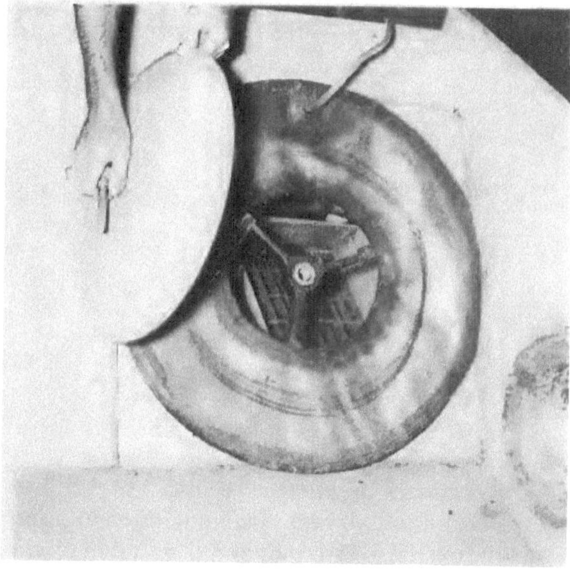

Figure 5.—Modified pneumatic O-ring for open ditch turnouts.

Lay-Flat Pneumatic Valve for Farm Ditch Turnout

The 15-inch lay-flat pneumatic valve, shown in figures 6-8, inclusive, is simply a flat rectangular, tube that inflates to form a closure within the underground part of the turnout pipe. It is clamped into position at the upstream end, and when deflated or in an open position it conforms to the curvature of the concrete pipe wall. As the valve inflates with air, it expands upward and outward until closure is complete and flow of water ceases. The time required to open and close the lay-flat valve is about 2 minutes, but can vary depending on the size of the valve and air restrictions in the pressure system.

Figure 6.—Lay-flat butyl rubber pneumatic valve, showing plastic tube connection and rubber flap to protect connection against damage from debris in water.

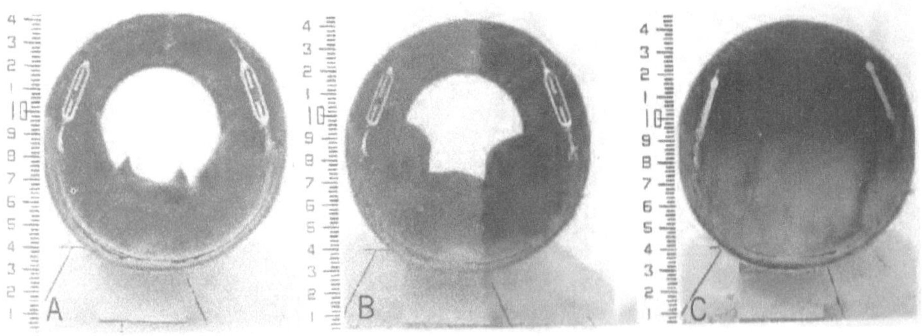

Figure 7.—Lay-flat 15-inch pneumatic valve of butyl rubber clamped in model pipe section to demonstrate three positions of valve: A, deflated or open: B, partially inflated; and C, inflated or closed.

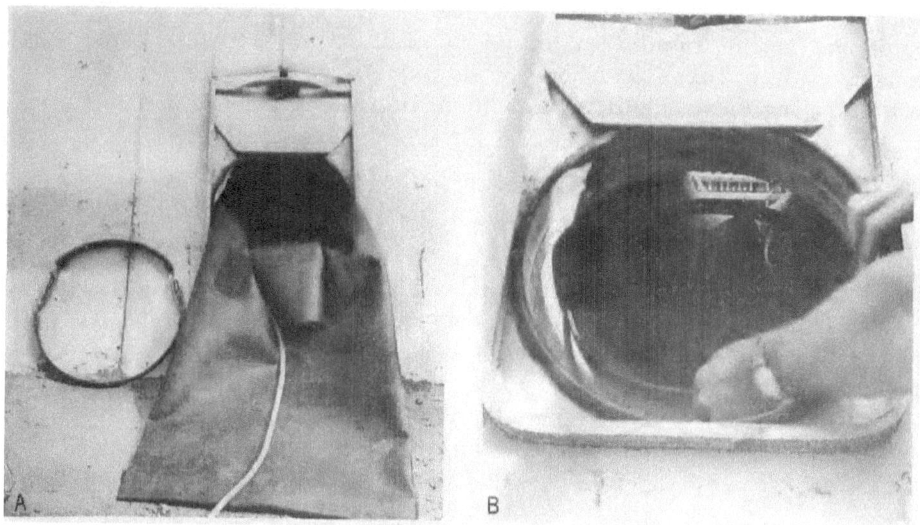

Figure 8.—Lay-flat valve in farm ditch turnout is easily installed. Note split clamp with turnbuckles (A) and in position at upstream end in turnout (B). Bellvue Outdoor Hydraulic Laboratory, Bellvue, Colo.

Valves have been constructed from both natural rubber and nylon-reinforced butyl rubber. The reinforced valve has greater strength, withstanding pressures up to 14.5 p.s.i. The natural rubber valve has a wall thickness of three-sixteenths inch and can withstand a pressure of only 3 p.s.i. Although the natural rubber valve has operated satisfactorily within its pressure limitations, the butyl rubber model offers a fivefold safety factor in case pressure regulators in the system fail to operate properly.

Pressure Supply and Control

Irrigation valves were operated under laboratory conditions with air and with water pressure. Valves opened and closed with air required about one-half the time of that needed for water.

Air under pressure is delivered to each valve or combination of valves by plastic pipelines commonly used in underground lawn sprinklers or from portable pressured tanks with pressure controls. Plastic pipelines costing from 3 to 5 cents per foot have been installed in the field on a limited scale. With 3/4-inch pipe, operation of valves with air pressured at 2 p.s.i. was successful at distances up to 1,000 feet. Head losses in the plastic pipeline are offset in part by the reservoir action of the pneumatic valves themselves. An open valve that closes between inflated valves on either side receives some air from the inflated valves at the same time that the air delivery system is being repressurized by a compressor located at a power source.

Use of a portable pressurized air tank with solenoid control valve and pressure regulator has been tested, but further study is required. Major air leaks in the system could deplete tank pressure in a short time and cause a malfunction of the automatic system. A minor leak in the pipe delivery system would cause the air compressor to run excessively, but the automatic system would continue to function. A major air leak would cause all pneumatic valves to open. '

The pressure regulator installed between the air compressor and plastic pipeline distribution system controls pressure in the 0- to 15-p.s.i. range, and at 3 p.s.i. will deliver 36 c.f.m. High flow with low pressure control is required for faster operation of the pneumatic valves.

A chain-type trencher has been used to place the plastic air lines at depths of 18 to 24 inches for protection against tillage operations, rodents, and direct exposure to the sun. Where several pneumatic valves are tied together by a common plastic air line for multiple operation, the plastic air line and the one from a central air pressure supply have been placed in the same trench. In one field installation the trench has been placed parallel to the farm lateral and near the discharge end of the farm turnout for convenience in connecting the pneumatic valves to the air lines.

The three-way solenoid control valve has two principal requirements: (1) Momentary energization to reduce battery power and (2) adequate air flow capacity of inlet and exhaust ports. The valve presently used has the momentarily energized feature, but is designed to operate under maximum pressures of 200 pounds. At low pressures, restrictions to air flow within the valve are unacceptable, particularly when the control valve opens or closes two or more lay-flat pneumatic valves. Closure time for three 15-inch pneumatic valves with the solenoid control valve in the system is about 5 minutes. Opening time is about 10 minutes. Without the control valve in the system, the same pneumatic valves can be opened or closed in less than 2 minutes.

Remote Control

Two remote control systems have been tested in the field. One employs signals transmitted by radio and the other by wire.

Components of the radio remote control system consist of (1) a 24-hour time clock for programing time intervals with an accuracy of 5 minutes, (2) a 12-channel citizens band transmitter capable of radiating 3 watts at a frequency of 27.235 mc. (fig. 9), and (3) 12 receivers, each of which is tuned to one of the transmitter channels (fig. 10). The time clock activates the transmitter automatically at the end of each interval.

The diagram of circuit and components used to switch consecutive impulses from one transmitter channel to the next is presented in figure 11. The system is programed so that one signal is transmitted to open the downstream pneumatic valve(s), and then 30 seconds later another signal is transmitted to close the upstream valve(s). Thus, one valve or set of valves is always open to receive the flow of water. Programing is flexible and any other sequence of operation can be obtained.

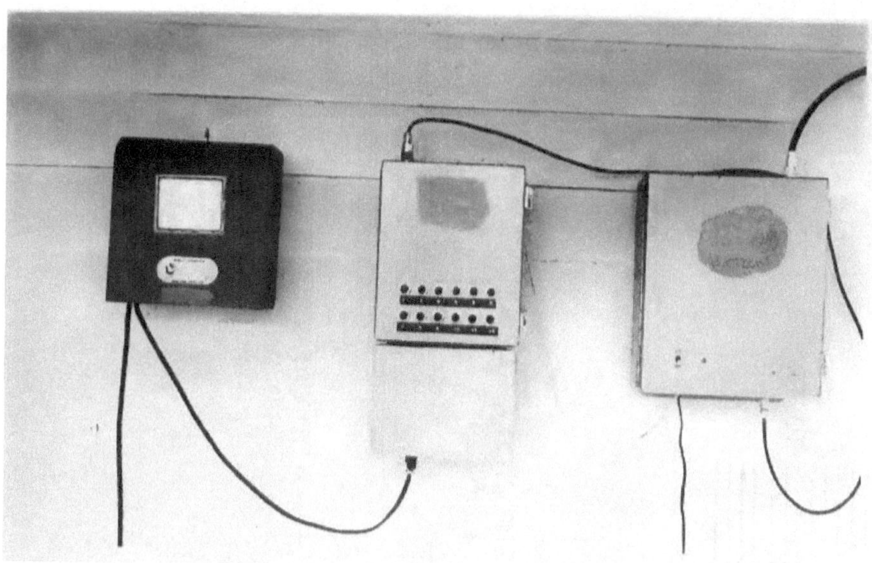

Figure 9.—Programing and transmission equipment. From left to right are a 24-hour timer, a 12-channel transmitter, and a power amplifier for the transmitter.

Figure 10.—Radio receiver and three-way solenoid air valve, with power supplies and carrying case.

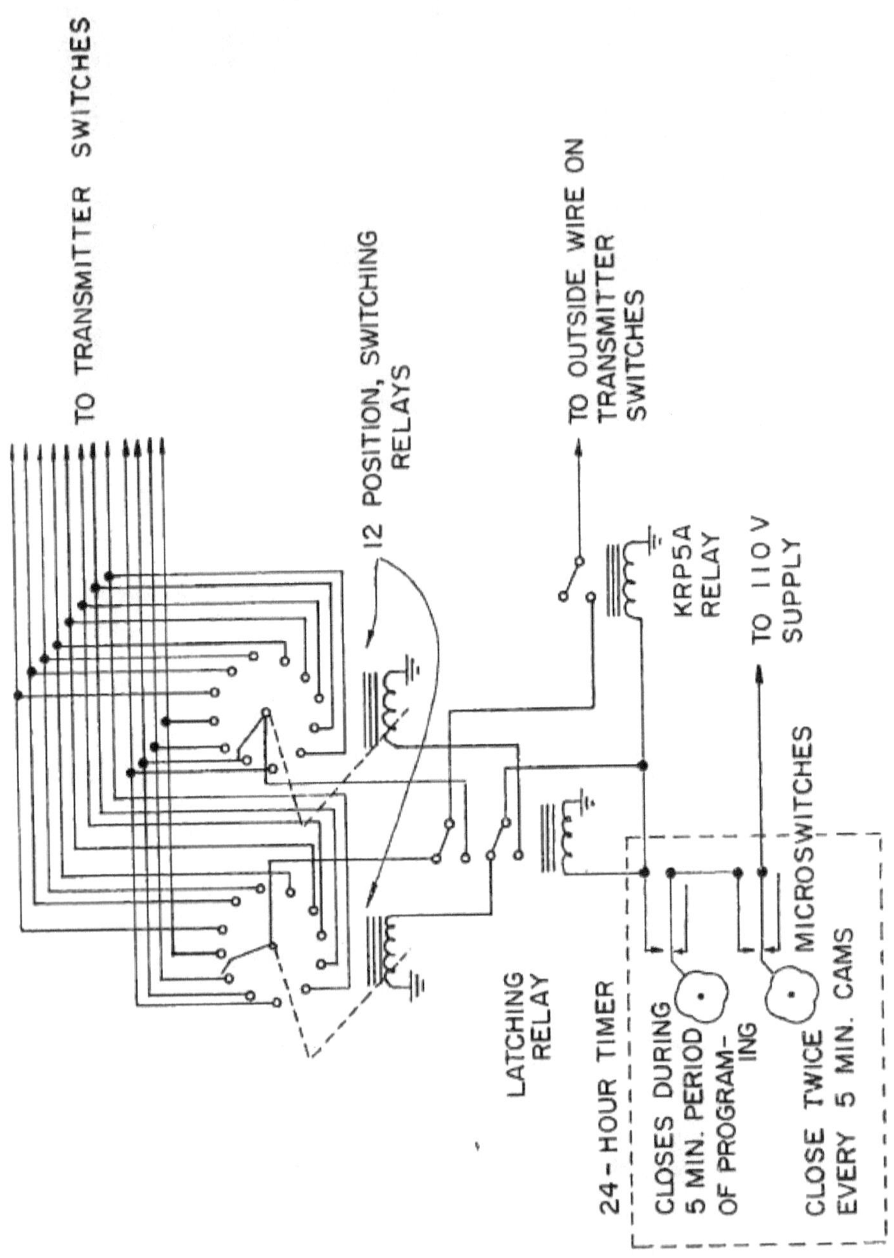

Figure 11.—Wiring diagram for timer and relay system used in conjunction with radio transmitter.

A supplementary power system is- required for the radio receiver to activate the 115-volt, momentarily energized, air control valve. A 67 1/2-volt dry cell with a capacitor is used for power to energize the solenoid. A latching relay switches consecutive impulses from one solenoid to the other so that the pneumatic valve is alternately opened and closed. The capacitor and relay are housed within the receiver case. A sketch of the wiring diagram for this supplement to the receiver is shown in figure 12. A control valve with 12-volt coils and larger inlet and exhaust ports is being developed for more efficient operation.

The industrial timer and components to open and close valves using wires instead of radio are shown in figure 13 and the wiring diagram is given in figure 14. One common wire and one for each of the control solenoid valves are required for each complete irrigation set. The wires can be enclosed in the same plastic pipeline used to convey air pressure to each pneumatic valve.

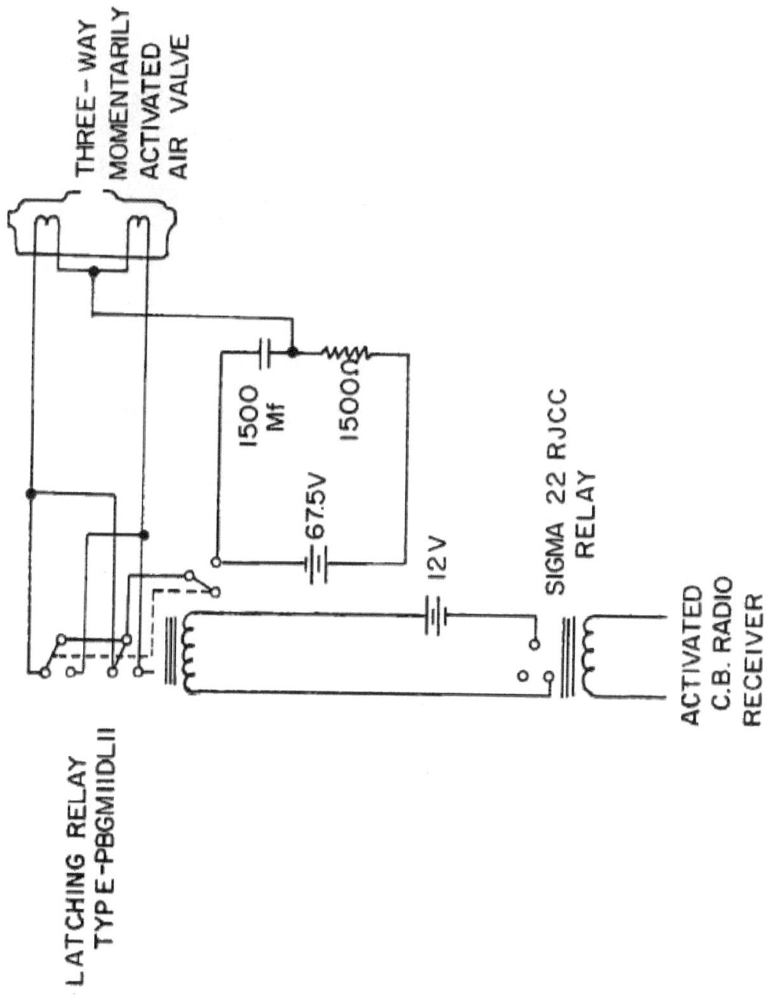

Figure 12.—Wiring diagram for power system used in conjunction with radio receivers.

Both types of remote control systems have been successfully tested in the field, but more experience is needed before the reliability of either system can be assessed and evaluated.

So-called off-the-shelf components were used to expedite testing of the remote control system described, regardless- of cost. A tone generator and discriminator system specifically designed to signal the pneumatic valves by a single pair of wires can be built for approximately 60 percent less than the commercial radio transmitter and receivers. Multifrequency signals transmitted by the pair of wires appear to offer the most practical method of control. Radio control requires a license, and when operated in a contiguous block of irrigated farms could conceivably result in unintentional operation of valves on an adjacent farm or farms.

Figure 13.—Timer and relay system for controlling pneumatic valves by wire.

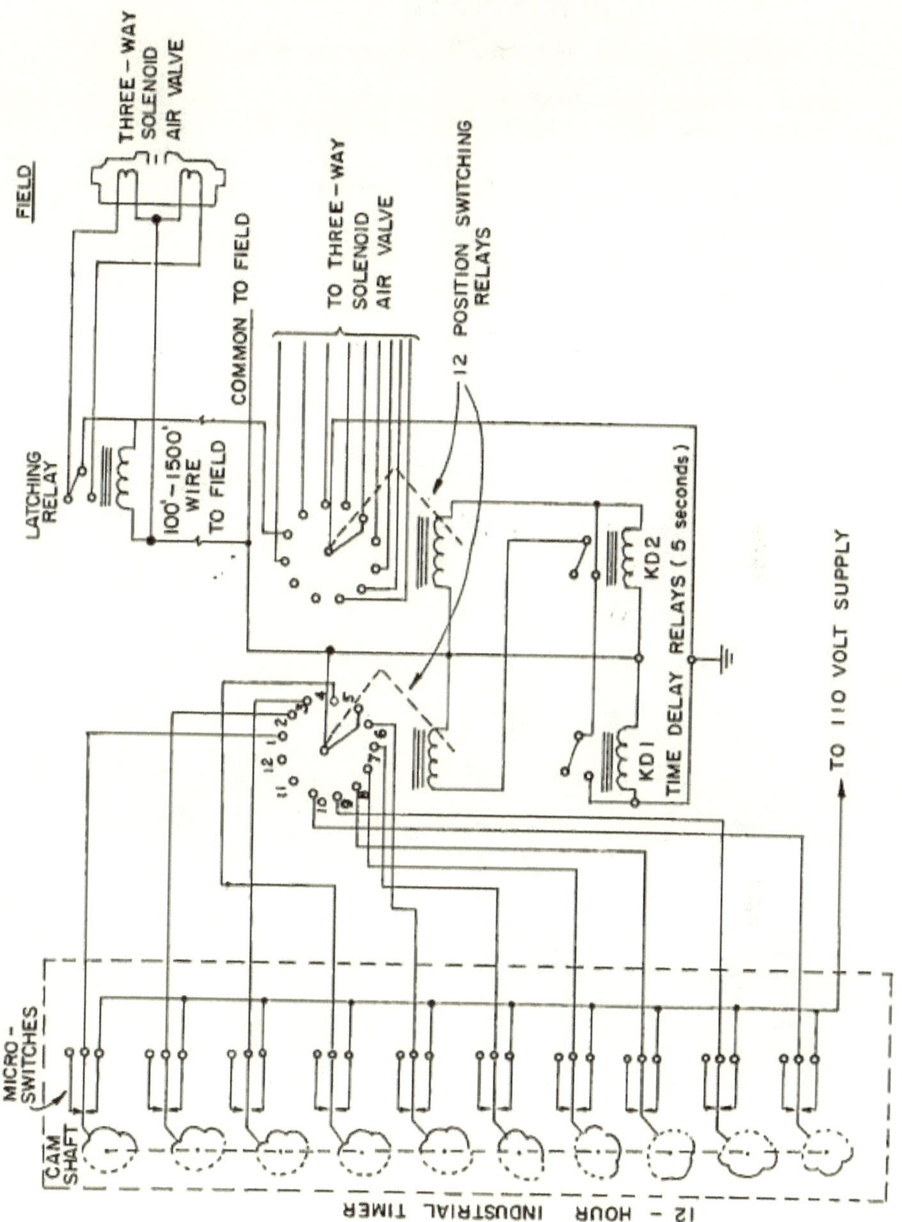

Figure 14. — Wiring diagram for timer and relay system used in conjunction with wires to each pneumatic valve.

FIELD INVESTIGATIONS

Pneumatic Valve for Pipe Irrigation System

A field installation of the automatic prototype was successful on level contour border strips of Pierre clay seeded to pasture at the Newell Irrigation and Dryland Field Station, Newell, S. Dak. (fig. 15). Radio signals were transmitted over a distance of one-half mile to open and close the pneumatic irrigation valve, previously described, according to a predetermined operating schedule. In the test, a stream of 3 c.f.s. was applied for four border strips of varying areas for periods of 30, 10, 15, and 20 minutes, respectively. The depth of water application was 3 inches on each border. To switch water from one border to the next, the downstream valve was opened first, and 30 seconds later the upstream valve was closed. All valves operated on schedule.

A minor erosion problem (fig. 16) was observed around the alfalfa valve in the early stages of opening. This has been avoided elsewhere by surrounding the alfalfa valve assembly with a concrete riser or adapting a hydrant for metal pipeline connections. The pneumatic valve for alfalfa valves has been used successfully in a hydrant attachment for furrow irrigation, but some modifications in design are required.

Figure 15.—Field installation of 12-inch pneumatic valve.

Figure 16.—Water flowing from automated 12-inch alfalfa valve. Protection is needed to prevent soil erosion from the turbulent, swift stream.

Modified Pneumatic Alfalfa Valve (O-Ring) for Farm Ditch Turnout

Limited field tests were conducted on the modified alfalfa valve at the Newell field station in South Dakota, but efforts to improve the design were abandoned in favor of the lay-flat model described previously. The valve in an open and closed position is shown in figures 17 and 18.

Although the modified pneumatic valve is effective in turning water on or off at the farm ditch turnout, there are several inherent disadvantages. Construction is complicated by the addition of spring steel, which increases the cost. Metal components adapted from the alfalfa valve are necessary for each turnout and are costly. Perhaps the greatest disadvantage is its susceptibility to trash problems and possible malfunction when unattended. Tumbleweeds and similar vegetation that might collect in the valve openings could reduce substantially the flow of water and prevent the valve from closing.

Figure 17.—Modified pneumatic O-ring in open position with water flowing. Note plastic air line and three-way solenoid control valve in background.

Figure 18.—Field installation of pneumatic irrigation valve. Valve in closed position. Installed in concrete farm lateral at Newell Irrigation and Dryland Field Station, Newell, S. Dak.

Lay-Flat Pneumatic Valve for Farm Ditch Turnout

Several 12-inch lay-flat valves were installed in turnouts at the Newell field station (fig. 19). The valves functioned well, causing no obstruction to the flow of water when open and stopping flow completely when closed. Only 2 p.s.i. of air pressure were required for operation.

Advantages of the lay-flat valve (fig. 8) are low overall cost, ease of installation, minimum obstruction to flow of water, minimum obstruction to trash, preadjustment of existing turnout gate for a given stream size, and ease in conversion of existing distribution systems with turnouts. Also, the valve is completely out of sight to would-be target hunters (fig. 19).

The lay-flat pneumatic valve has been constructed and successfully tested in 12- and 15-inch concrete pipe turnouts. Commercial valves are available in these sizes and can be produced in other sizes if desired.

Figure 19.—Twelve-inch lay-flat pneumatic valve in open ditch turnout.
Note unobstructed flow of water.

FLOW RATES THROUGH PNEUMATIC IRRIGATION VALVES

Manual controls, either alfalfa valves on pipelines or slide gates in open ditches, provide a means for adjusting the rate of flow and also a positive shutoff if the automatic system fails. Rates of flow were measured through several gates with pneumatic valves in the laboratory. These flow measurements serve as a guide to the proper setting of the gates in the field so that the desired flow rates may be obtained. However, conditions in the laboratory and field, such as backwater on turnouts and points of head measurement, may differ, and the following data may need adjustment before being applied to the field.

Flow rates were measured with pneumatic valves in place in the following types of turnouts: (1) 12-inch alfalfa valve discharging freely into the atmosphere; (2) 12-inch alfalfa valve surrounded by a cylinder that caused submergence of the flow; (3) 3-foot horizontal section of 12-inch pipe containing a 12-inch lay-flat pneumatic valve—flow was unsubmerged and head was taken as the difference between the water surface in the ditch and the invert of the pipe; and (4) a 15-inch slide gate and concrete pipe turnout containing a 15-inch lay-flat pneumatic valve—the invert of the slide gate opening was 4 inches above the invert of the ditch and flow from the pipe was unsubmerged.

Unsubmerged discharge through the alfalfa valve was measured and related to the valve opening expressed as number of turns from the closed position. Sketches showing installation of valves, flow conditions, and points of head measurement are presented in figure 20. The thread pitch on the screw supporting the valve lid is such that three complete turns equal 1 inch. The data obtained on the flow rate are shown in figure 21. A maximum discharge of 4.6 c.f.s. was obtained through this valve with the lid nine turns (3 inches) open and a piezometric head of 1.46 feet. The pneumatic valve was capable of shutting flow off completely at this opening.

If a structure is placed around the alfalfa valve to prevent erosion, outflow from the valve is likely to be submerged. The effective head causing flow will be the difference between the piezometric head in the pipeline and the water surface elevation at the valve, both referenced to the same datum. Flow rates measured for such conditions are shown in figure 22. Maximum discharge was 4.75 c.f.s. for a gate opening of 10 turns and a differential head of 1.0 foot.

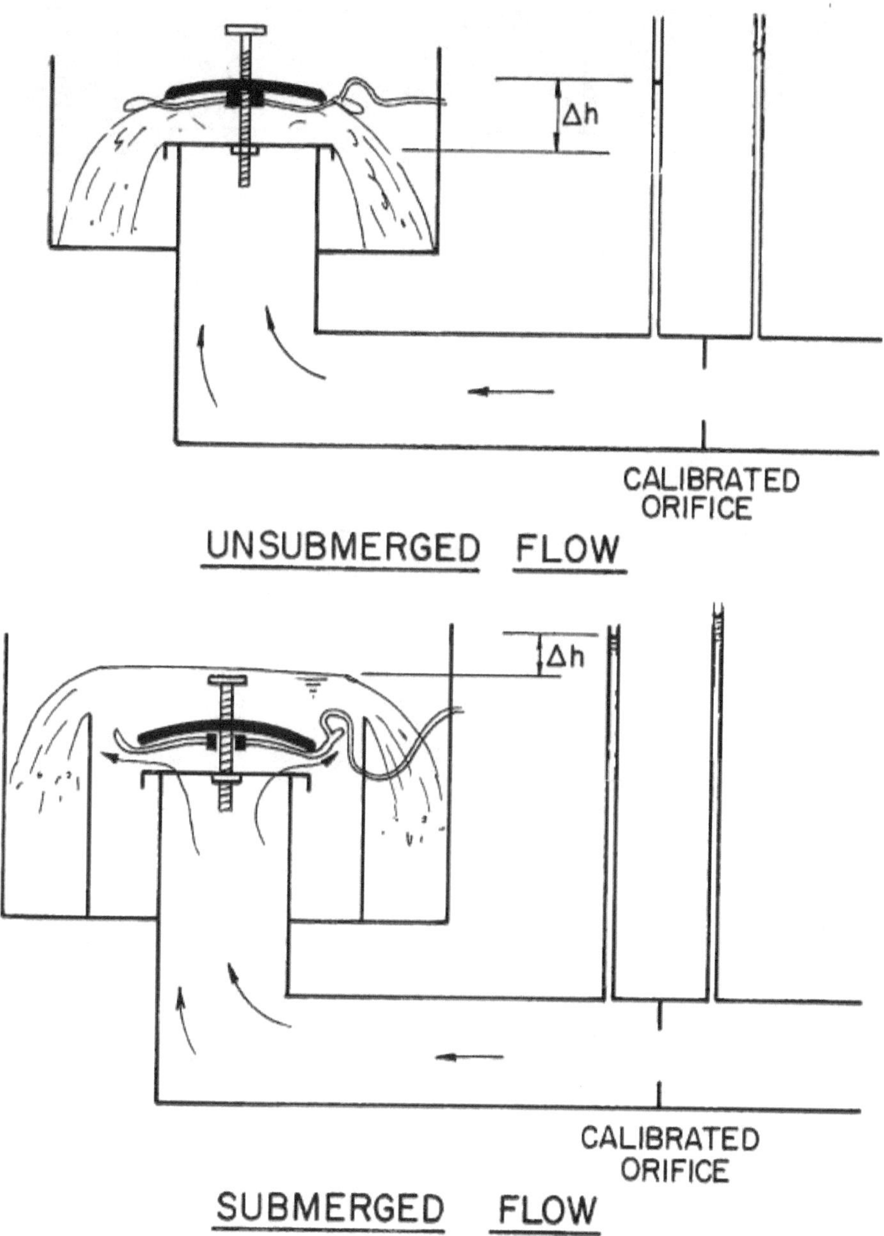

Figure 20.—Sketch of flow conditions during alfalfa valve calibration.

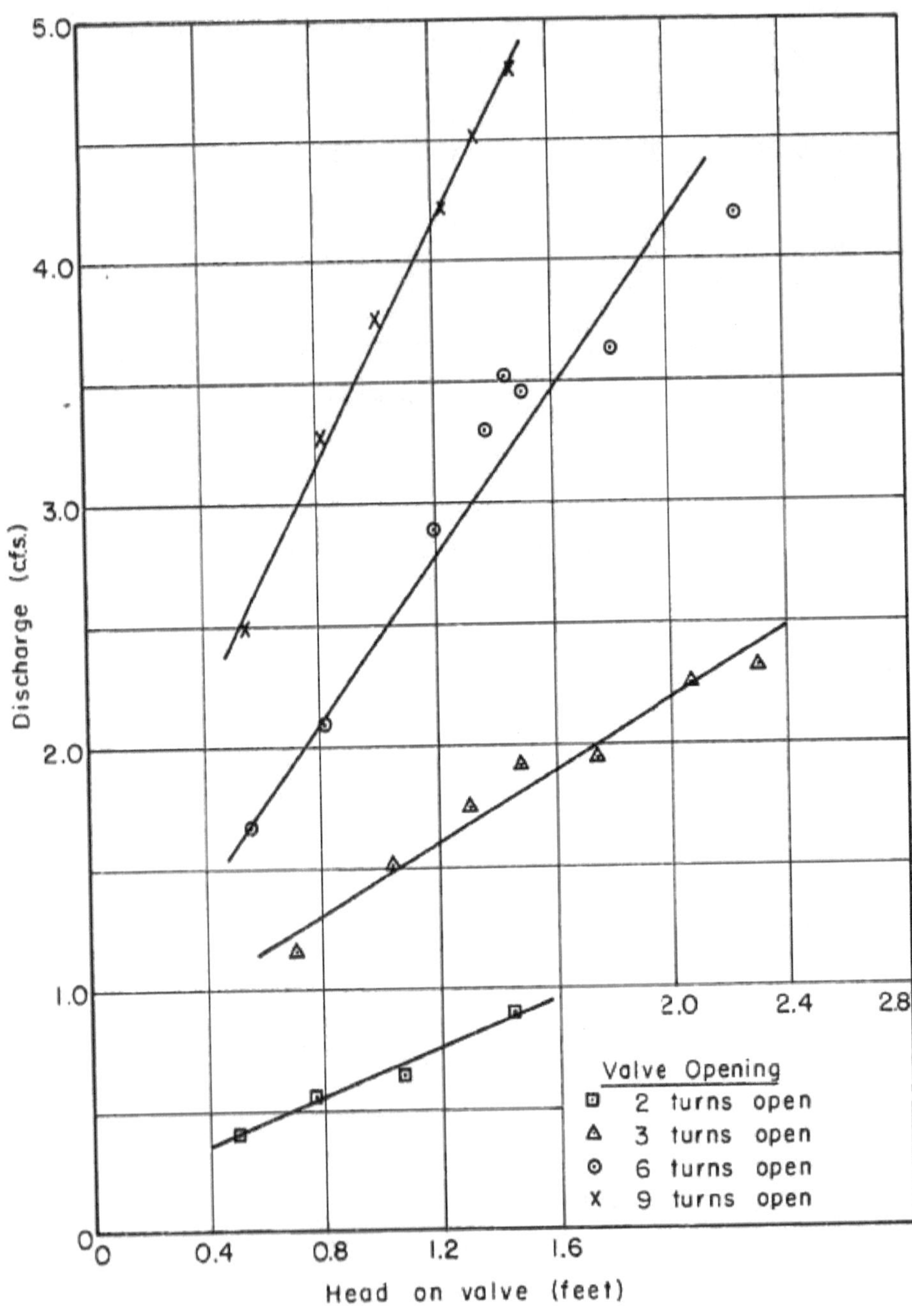

Figure 21.—Discharge through unsubmerged 12-inch alfalfa valve with pneumatic valve installed.

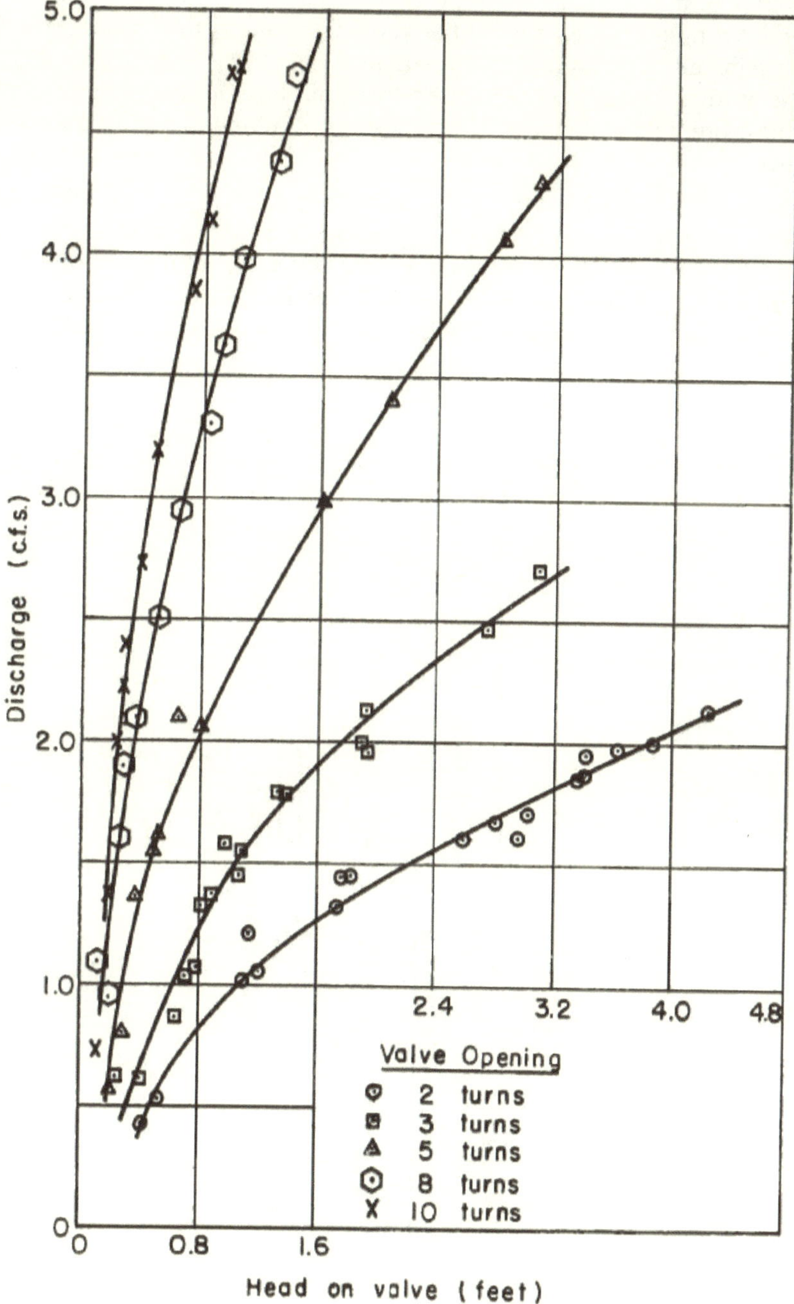

Figure 22.—Submerged discharge through 12-inch alfalfa valve with pneumatic valve installed.

The rate of flow through the pipe of a 12-inch open ditch turnout was measured both with and without a lay-flat valve installed in the turnout. Discharge was not reduced by installing the pneumatic valve. The flow obtained through this type of turnout was 2.63 c.f.s. when the water level in the ditch was 0.80 foot above the center line of the pipe. The pipe was not submerged during these measurements.

Flow was also measured through a 15-inch open ditch turnout with a lay-flat pneumatic valve in place. A sketch of the slide gate, concrete pipe, and pneumatic valve is shown in figure 23. Maximum flow measured through this turnout was 3.9 c.f.s. with a depth of flow in the ditch of 1.90 feet (fig. 24). With the same depth of flow and the turnout gate one-half open, the flow is reduced to 2.58 c.f.s.

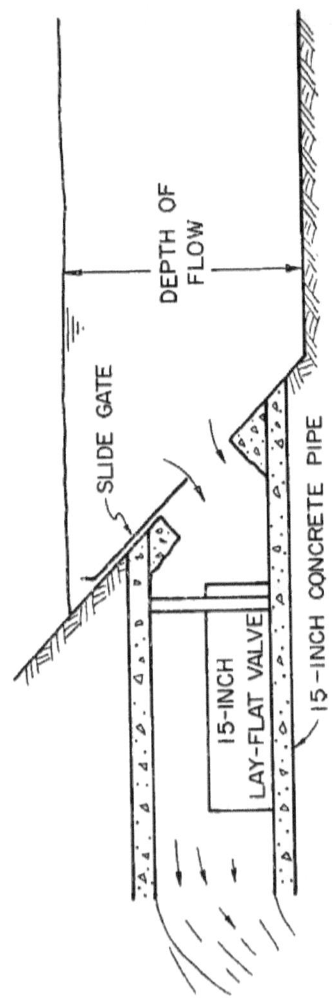

Figure 23.—Sketch of flow conditions during calibration of 15-inch open ditch turnout.

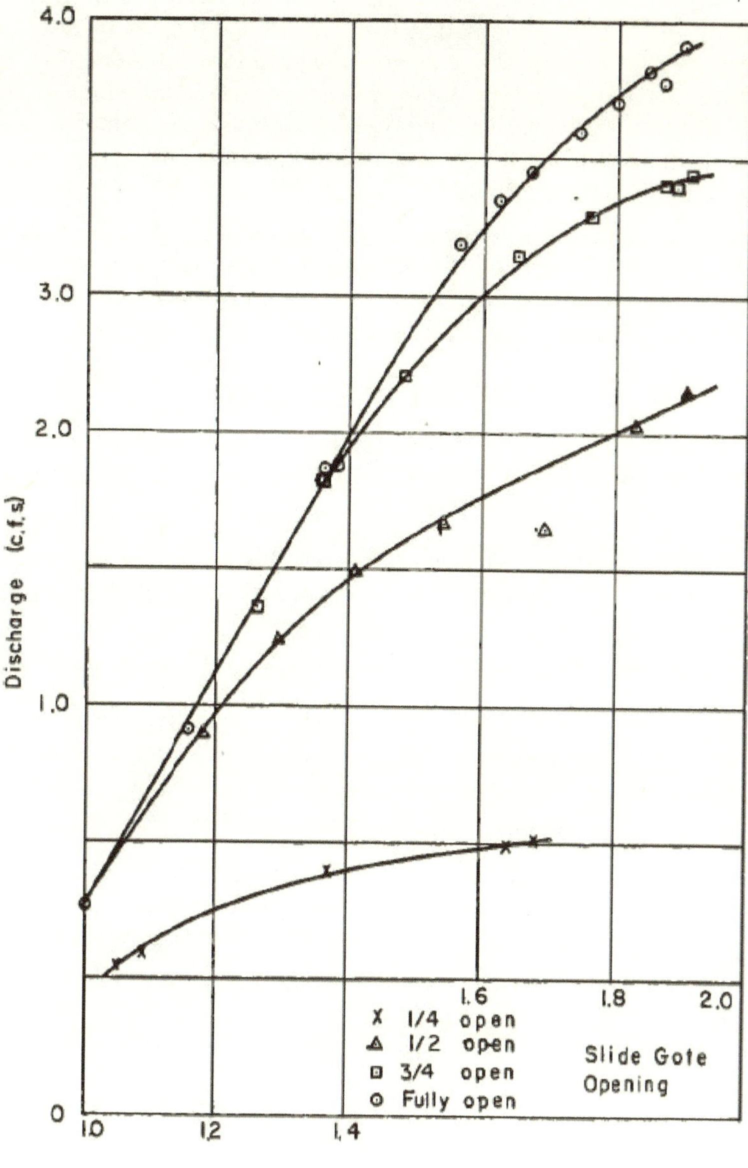

Figure 24. — Discharge through 15-inch open ditch turnout.

Recommendations

When the pneumatic valves used to control flow through alfalfa valves are deflated, the pressure of the water forces them up against the valve lid. Thus, the interference to flow is very small. If the valve opening is measured in terms of number of turns from a fully closed position, the effective opening should be the same with or without the pneumatic valve. Therefore, on existing systems where operators through experience have learned the optimum valve opening, the same opening should be used after pneumatic valves are installed. Similarly, laboratory tests showed that the lay-flat valves provide negligible resistance in open ditch turnouts. The same slide gate openings used for manual operation should be used after pneumatic valves are installed.

When pneumatic valves are installed in a new system, gates should be set to give the flow rates specified by the design engineers, using information in this section as a guide to gate openings. Adjustments may be necessary after observing the functioning of the system during the first one or two irrigations.

FUTURE RESEARCH

Further studies on the development and testing of the pneumatic irrigation valve should include—

(1) Comparative farm-scale studies of such items as labor costs and irrigation application efficiency, with and without automation.

(2) Automation of farm headgates to permit remote regulation of project irrigation distribution systems.

(3) Development of "fail-safe" features if automated system malfunctions.

(4) Development of portable pneumatic checks for farm laterals.

(5) Development of specific components to improve overall operation of automated system.

(6) Tests of durability of pneumatic valves.

(7) Exploration of possibilities of using sensing elements to actuate control valves on the basis of such factors as soil moisture depletion or replenishment and advance of water down the border, with the view toward complete automation.

SUMMARY

Several models of a pneumatic nylon-reinforced valve for remote control of water application from underground pipeline or open ditch distribution systems are described. Basic components of the system include (1) a pneumatic closure, (2) a three-way solenoid valve that permits the flow of air into or out of the pneumatic closure, (3) a source of air pressure to inflate the closure, and (4) a centrally located remote control system with timing device to actuate the three-way solenoid control valve by means of a signal transmitted by radio or carried by wire. Preliminary results of field tests at Newell, S. Dak., demonstrated the feasibility of remotely controlling water applications on farm fields by radio up to distances of one-half mile.

Lector House believes that a society develops through a two-fold approach of continuous learning and adaptation, which is derived from the study of classic literary works spread across the historic timeline of literature records. Therefore, we aim at reviving, repairing and redeveloping all those inaccessible or damaged but historically as well as culturally important literature across subjects so that the future generations may have an opportunity to study and learn from past works to embark upon a journey of creating a better future.

This book is a result of an effort made by Lector House towards making a contribution to the preservation and repair of original ancient works which might hold historical significance to the approach of continuous learning across subjects.

HAPPY READING & LEARNING!

LECTOR HOUSE LLP
E-MAIL: lectorpublishing@gmail.com

www.ingramcontent.com/pod-product-compliance
Lightning Source LLC
Chambersburg PA
CBHW030420160726
47992CB00007B/3216